iPhone 12 Pro and iPhone 12 Pro Max User Manual
For the Elderly

An Illustrated Step By Step Guide with Tips and Tricks to Operate the New iPhone 12 mini, iPhone 12 Pro and iPhone 12 Pro Max For Seniors

DERRICK
RICHARD

Copyright

Contents

CHAPTER ONE

SETTING UP YOUR iPhone 12

Upon purchase and delivery of your new iPhone 12, you will surely un-box the device from the box to find your new iPhone with a USB Type C cable. The next thing is to turn on your device, configure some basic settings and start to use the phone. For proper setting up of your device, endeavor to put the following tips into consideration;

- Your iPhone 12 will be configured over a secured cellular data for the internet. If a cellular service is not available, you can use Wi-Fi to set up your phone.
- Your Apple ID and Password. You don't need to set up an apple ID and password if you already have one. But if you don't have an Apple ID and password, you can register during the initial set up of your device.
- A backup for your iPhone. You can use your old iPhone or iPad as a backup. Transferring files from your old iPhone or iPad to the new iPhone 12 will require you to set up a backup to prevent file loss in the process.
- An Android phone. This is when you want to transfer your old files from an Android device to your new iPhone 12.

How to turn on and set up your iPhone 12

- **Use the side button to turn on your iPhone 12:** iPhone 12, unlike some iPhone models with sleep/wake button at the top of the device, has the side button, which will be used to power the phone ON. Bring up the Apple logo (power on) by pressing and holding on the side button. On the iPhone 12, the side button is found to the left side of the phone.
- **Set up your iPhone manually or by using the Quick Start option:** Any of the method below can be used to set up your iPhone 12;
  - Select the "**set up manually**" option once you see the Apple logo coming up on your phone and you will see the on-screen instructions to guide you.
  - The **Quick Start** menu can as well be used to set up your new iPhone 12. This method will be used if you have another Apple device (possibly a

device running iOS 11 or higher or iPadOS 13 or higher). To use the Quick Start option, you will need to copy some of your device settings, iCloud and preferences by bringing your new iPhone 12 in close contact with another Apple device containing your settings. The Quick Start option ensures that your former settings on your old iPhone are retained on your new iPhone 12. Although the Quick Start option might not be able to transfer all of your old settings, you can always copy the rest of the settings with your iCloud backup.

To use the **Quick Start,** follow the instructions below;

1. Power on your new iPhone 12 (use the side button) and bring the iPhone 12 in close proximity with your old iPhone (containing all of the settings you want to transfer to the new iPhone). Once the two

devices are in close range, you will receive a Quick Start screen on the old device with a directive to use your Apple ID for setting up your phone. Be sure it is your Apple ID displaying on your old gadget. Once the Apple ID has been confirmed, simply tap on **continue.** If the link to continue is not showing, try and turn on your old device's Bluetooth.

2. Wait a little while till you receive an animation on the iPhone 12 screen. Gently position your old iPhone on top of the iPhone 12 and allow the animation to center appropriately in the viewfinder. You will get a prompt that says **Finish on New.** Manual authentication can be done in case you are unable to use the phone camera at that time.

3. When requested, enter the passcode for your old iPhone on the new iPhone 12.

4. Proceed with the on-screen guide to set up your Touch ID or Face ID on the new iPhone 12.

5. When requested, fill in your Apple ID and password on your new iPhone 12.

6. You will get an option to transfer your files or to restore apps, data and settings from the iCloud backup. You can equally tap on **Other Options** if you want to restore data from a backup on your personal computer. Once you choose backup, transferring settings related to location, Apple pay and Siri becomes easier.

7. Also, for users with an Apple Watch (say Apple Watch 5 or Apple Watch 6), you can be asked to transfer the contents on your Apple Watch during set up.

You can also use device-to-device migration to transfer your contents;

- Power on your new iPhone 12 (use the side button) and bring the iPhone 12 in close proximity with your old iPhone (containing all of the settings you want to transfer to the new iPhone). Once the two devices are in close range, you will receive a Quick Start screen on the old device with a directive to use your Apple ID for setting up your phone. Be sure it is your Apple ID displaying on your old gadget. Once the Apple ID has been confirmed, simply tap on **continue.** If the link to continue is not showing, try and turn on your old device's Bluetooth.

- Wait a little while until you receive an animation on the iPhone 12 screen. Gently position your old iPhone on top of the iPhone 12 and allow the animation to center appropriately in the viewfinder. You will get a prompt that says **Finish on New.** Manual authentication can be done in case you are unable to use the phone camera at that time.
- Proceed with the on-screen guide to set up your Touch ID or Face ID on the new iPhone 12.
- When requested, fill in your Apple ID and password on your new iPhone 12.
- Tap on "Transfer from [name of your old iPhone]" to begin to transfer your contents/files from your old iPhone to the newly bought iPhone. You will see the

icon showing you that the two devices (new phone and old phone) have been properly connected if you are using the wired method to transfer your files. You can equally decide to transfer some of your previous device's settings like the Siri settings and Apple Pay settings.

- Also, for users with an Apple Watch (say Apple Watch 5 or Apple Watch 6), you can be asked to transfer the contents on your Apple Watch during set up.

Make sure that both the new and old phones are placed near each other and that they are properly plugged for charging until the data transfer process is completed. After completing the set up, you might be asked to configure further settings for your device. Most of these settings will be discussed in the latter part of this guide.

Setting cellular service on iPhone 12

An eSIM or a Nano SIM is required for some of the cellular functions on your iPhone 12. The iPhone 12 supports a dual SIM meaning that you will be able to use an eSIM together with a Nano SIM on your phone. With either an eSIM or a Nano SIM, you can: send and accept calls, send and receive messages and also access the internet. With both type of SIMs supported on the iPhone 12, you can;

- Use one of the SIMs for voice call and the other SIM for your data plan.
- Use one of the SIMs for your business and the other SIM for your personal assignments.

How to install Nano SIM on your iPhone 12

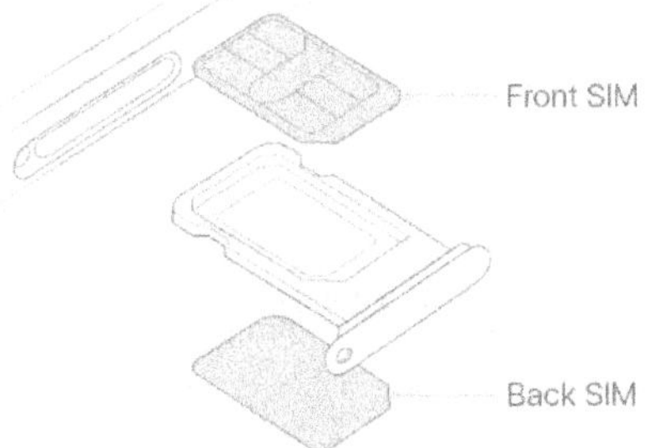

- Older iPhone models usually come with a SIM ejector, but in case your new iPhone 12 does not have a SIM ejector, you can always use a paper clip or a pin that can enter inside the small hole of the SIM tray.

- Remove the SIM tray on the side of your device by inserting the SIM ejector inside the small hole. Be careful while removing the SIM tray to avoid accidental damage of the tray.
- Place your SIM card on the SIM tray with care.
- Place the SIM tray (with the SIM now inserted) into the hole while ensuring that the SIM tray enters the hole fully without leaving any crevices. This is to make sure that no liquid or dirt can enter the hole.

Setting up a SIM PIN for your iPhone

Setting up a SIM PIN for your new iPhone 12 allows you to secure your SIM card from unauthorized access by anyone (in case of theft, maybe) who might have gained unauthorized access to your iPhone 12. But be reminded that setting up a SIM PIN for your SIM card will always require you to input the SIM PIN anytime you remove your SIM from the SIM slot and put it back. To set up a SIM PIN for your iPhone, follow the steps below;

- Navigate to the **settings** app on your iPhone 12, click on **Cellular** and select **SIM PIN.**
- Toggle on the SIM PIN menu and you will be prompted with a screen where you can input a secured SIM PIN for your iPhone 12.
- In case you have never set up SIM PIN for that particular SIM card before, simply enter the default PIN from your SIM carrier. Do not try to guess the default SIM PIN if you don't know the SIM PIN as this will usually get your SIM

blocked. Contact your SIM carrier if you don't know your default SIM PIN.
- Tap on "**Done**" to finish.

Setting up cellular plans with eSIM

Your iPhone 12 supports eSIM and a physical SIM (Nano SIM). The eSIM is stored digitally on your iPhone 12. To set up cellular plan with eSIM, follow the guides below;

1. Scroll to the **Settings** app on your device, select **Cellular** and then tap on "**Add Cellular Plan**."

2. The QR code assigned to you by the service provider must show in the camera frame when you properly position your iPhone. You can also input the details manually. You may be requested to input a confirmation code from your carrier.

3. Tap on "**Add Cellular Plan**."

Although, the iPhone 12 allows users to have more than one eSIM; but you can only use one eSIM at a time. You can always switch between eSIM by going to **Settings,** tap

Cellular, choose the plan you want to use and then select **"Turn on this line."**

Manage your cellular plans

Since your iPhone 12 supports Dual SIM (eSIM and Nano SIM); you can choose how you want your iPhone to be using each of the SIM. This means that you can use one SIM for call and the second SIM for data plan. To do this;

1. Scroll to **Settings** app on your device and tap on "**Cellular.**"

2. Do the actions below;

   - Tap on **Cellular Data** and then choose a default line. Enable the "**Allow Cellular Data Switching**" to let your iPhone 12 use either of the two lines depending on network availability and coverage. Roaming charges will always be incurred on your lines if you activate **Data roaming** and you are not within the region that your carrier network covers.

- Click on "**Default Voice Line**" and then choose a line.

- Below Cellular Plans, pick a line and proceed to change settings for the line. Settings such as Cellular Plan Label, SIM PIN, Calls on other devices and Wi-Fi Calling (if your carrier enables it) can be changed for the SIM.

Kindly remember the following if you are using Dual SIM;

- If you are currently utilizing one of your lines to receive a call, **Wi-Fi calling** must be activated for the second line if you still want to receive calls on it. In any instance that a call comes into your second line while your first line is currently in use, and you don't have an active Wi-Fi connection, your iPhone will utilize the cellular data of that particular line you are currently using to call to accept the call on the other line. Charges may be applicable. You will have to enable data use in the cellular settings for the line that you are currently using to be able to take calls on the other line.

- If you don't activate Wi-Fi Calling for a particular line, any call that comes on that line (not excluding calls from emergency numbers) will be taken to voicemail directly (if voicemail is available for your carrier) when you are currently using the other line and you won't be able to receive missed call notifications. Setting up conditional call forwarding from any of your mobile lines to another line when you are busy using the line or not in service will not direct your call to voicemail.

- Your default line will be activated for use whenever you send a call from some other devices, such as personal computers or Mac. The call will be sent through your iPhone 12 with Dual SIM.

- Switching chats from a line you initiated SMS/MMS with to another line on your device will not be possible. This means that the line you are using to type an SMS is the exact line you will have to use to send the chat. If you want to switch lines at all, you will need to erase the entire chat and then begin another chat all over again with the second line. You need to also know that extra charges will be involved when sending an SMS/MMS attachment on any of the lines you are not using for cellular data. The line you selected for your cellular data is the line that will be deployed for Instant Hotspot and Personal Hotspot.

Enjoying the internet on your iPhone 12

Connecting the iPhone 12 to the internet can either be done using an opened Wi-Fi connection or your secured cellular service.

Connecting iPhone 12 to a Wi-Fi network

1. Scroll to your phone's **Settings**, click on **Wi-Fi** and then toggle on the Wi-Fi.

2. Tap on one of the options below;

- *A network:* You might be prompted to fill in a password.

- *Other:* Choosing this will let you connect to a hidden network. Fill in the name of the network, password for the network and the security type for the network.

A successful connection to a Wi-Fi network will bring the Wi-Fi symbol at the top section of your notification screen. To ascertain this, simply go to the Safari web browser and your web pages will begin to load automatically.

Join a Personal Hotspot

You can have your friend (with an iPad that has Wi-Fi + cellular or an iPhone) share his/her personal cellular connection so that you can enjoy internet connections if you don't have active data on your own device. Follow the guides below to enjoy your friends' personal hotspot;

- Scroll to **Settings**, tap on **Wi-Fi** and then click on the name of the device that wants to share their personal hotspot. Some friends can use their names as the name of their hotspot.

- If you are requested to fill in the password for the hotspot, quickly ask the one sharing the hotspot to give you the password. In case the person does not know their password offhand, ask him/her to visit the **settings app** on their iPhone, select **Cellular** and choose **Personal Hotspot** to see their hotspot's password. If the password is hidden, they can tap on the **eye** symbol to make it visible.

If your friend has an Android device, he or she can join your iPhone's personal hotspot. Although, this requires some tricks around. Get started by downloading and installing a QR code generator that can create a QR code according to your device's Wi-Fi setting. This QR code will be scanned by the Android device and then applied to join the network. Follow the steps below;

1. Locate the Wi-Fi settings of the network. You will need the network's public name, password for the Wi-Fi and the category of wireless security (like WEP, WPA or none).

2. On your iPhone 12, download and then install a QR code generator capable of creating codes according to the settings of your Wi-Fi. There are many great apps you can use to generate this QR code. For instance, you can use **Visual Codes, which** is hugely recommended for saving multiple Wi-Fi QR codes on your device for use later. To use the Visual Codes app, download it on https://apps. apple.com/us/app/visualcodes/id12628 82475, and follow the steps from number 3.

3. Launch the Visual Codes app on your iPhone 12.

4. Click on **Add Codes**.

5. Select "**Connect to Wi-Fi**" at the lower end of the screen.

6. Enter the network's name inside the **Name** field.

7. Input the password for the Wi-Fi, and choose the correct security type (it is usually WPA here).
8. Enter a name that you will not forget for this Wi-Fi connection in the **Label** field.
9. Click on **Create Code**. You will get your new code on the Codes page, as part of a list alongside any other codes you have created for some other Wi-Fi networks. Choose an entry to bring its QR code.
10. On your Android gadget, launch the camera and position the Android phone so that it will be able to scan the code with the camera.
11. When you see the Wi-Fi network pop-up message, tap on it to connect your phone automatically to the network.

Connecting your iPhone 12 to a secured cellular network

If your iPhone 12 cannot identify any available Wi-Fi network to connect with, your device will connect automatically to your SIM card's cellular network. In case your iPhone 12 is unable to establish a connection, troubleshoot by;

1. Making sure that your SIM card is not locked with a passcode and that your SIM card has been activated.

2. Scroll to **Settings** and click on "**Cellular.**"

3. Make sure that your cellular data is turned ON.

Anytime your iPhone 12 needs an internet connection, the following actions will be done accordingly until it finds a secured connection;

- iPhone 12 will look for a Wi-Fi you used recently, and once connection is available; your device will connect with the Wi-Fi.

- Your device will get you the list of Wi-Fi networks close to you and then connect with the one you choose.

- Connects to your carrier's cellular network.

Note: If the iPhone 12 is unable to find a secured Wi-Fi connection within close proximity to you, your apps and services may try to send data using your carrier's cellular network. This may incur extra charges. Your SIM carrier has the cellular data rate and you can consult them for available data plans.

How to manage your Apple ID and iCloud settings on iPhone

Apple services such as your iTunes Store, App Store, Apple Music, iCloud, FaceTime, Apple Book, iMessage and a host of other services can be accessed with your Apple ID account.

Your apps, music, videos and documents will remain updated across all of your devices when you store them on the iCloud. With Apple iCloud, you can also share calendars, locations, photos, and a lot of other things with your friends and families. iCloud can also be used to find and locate your iPhone 12 if you misplace it or lose it to theft.

With iCloud, you will have access to a free email account with 5GB of space that can store your files, mails, photos and backups. Although, you will be able to upgrade your iCloud to get more spaces for file storage.

Sign in with your Apple ID

If you could not sign in to iCloud during the initial setup of your phone, you can do these;

1. Scroll to **Settings** .

2. Tap on **Sign in to your iPhone**.

3. Fill in your Apple ID and password. You will be able to create an Apple ID and password if you don't have one before. Also, you can recover your Apple ID and password if you forget them.

4. If you are using two-factor authentication for account security, you will be requested to input the six-digit verification code sent to your device.

Change your Apple ID settings

1. Scroll to Settings and tap on **your name.**

2. You will be able to carry out any of the actions below;

   - Update your device's contact information

   - File for another password.

   - Manage your device's Family Sharing

Change your iCloud settings

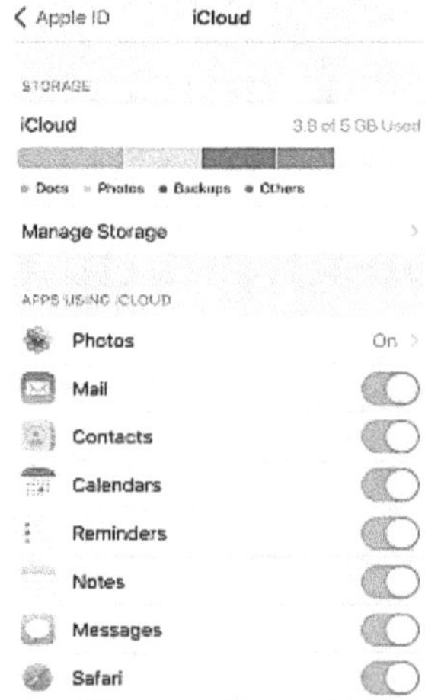

1. Select the **Settings** app on your device, click on **your name** *and tap on* **iCloud**.

2. Here, you will be able to do any or all of
   these;

   - See and monitor your iCloud's storage
     status.

   - Upgrade the iCloud storage status. To
     upgrade your iPhone's iCloud storage
     status, click on "**Manage Storage**" and
     select "**Change Storage Plan**."

   - Enable the features that you want to
     use, Photos, Contacts, Mail and Mess-
     ages.

Ways you can use iCloud on your iPhone

12

The following features can be updated with
iCloud;

- Contacts, Messages, Mails, Reminders,
  Calendars and Notes.

- Your images and videos.

- Apple Music, Apple Books and apps.

- Documents

- Your browser's Bookmarks, reading list, and web pages you have opened in Safari;

- Credit cards and Passwords.

You will also be able to carry out the listed actions;

- See and manage your iCloud data on your iPhone.

- Share your pictures and videos with individuals, friends and families you chose.

- You can also share your iCloud storage plan with about five (5) other family members or individuals provided you are on a storage plan with 200GB or higher.

- Locate your family members, friends or even your own missing devices, such as Apple Watch, AirPods, Mac, iPod touch, or iPad.

- Send live locations to your families and friends. This lets you locate any of them easily.

CHAPTER TWO

How to wake and unlock your iPhone

Conserving power is one of the in-built designs of your iPhone 12. Your device will normally enter sleeping mode and turns off its display anytime you are not using your phone. If your phone's display has been turned off or you want to exit the sleeping mode, you can wake your phone by using any of the method described below;

- Your iPhone 12 doesn't have the sleep and wake button at the head of the device, but it does have a side button. Simply press the **side button** (check the side of your phone to see the side button) to bring up your phone's display again or to bring it back from the sleep mode.

- The **raise to wake** feature on iPhone 12 will normally light up your phone's screen by merely lifting your phone to see it. This allows you to wake your iPhone 12 by raising your device. This feature can be disabled by tapping on **settings** and click on **Display & Brightness.**

- A gentle tap on your phone's screen can also wake your device. This is called the **tap to wake** feature.

Unlocking iPhone with Face ID

Leverage the Face ID feature to improve security for your device. With the Face ID, you have a good and trusted means of unlocking your device, purchasing and making instant payment, and also sign in to some third party apps on your phone. Face ID entails you to position your face at the right place on your phone. Before using Face ID, you will have to set up the Face ID feature. To set up the Face ID feature, follow the steps below;

- Set up passcode for your phone;

  o Your iPhone 12 comes with the Face ID feature. From **settings,** tap on **Face ID & Passcode**. There is no room to use the Touch ID & Passcode option in iPhone 12 unlike some old iPhone with the Touch ID & Passcode option.

- Set up the **Face ID** by;

   o Scroll to **Settings** , click on **Face ID & Passcode** and tap on **Set up Face ID.** You will be prompted with some on-screen instructions for full set up of the Face ID feature.

   o You will be able to add extra face appearance to complement the one you have added. The second face you added in this case will also be able to unlock your device. To get this done, scroll to **Settings** , tap on **Face ID & Passcode** and click on **Set up an alternate face appearance.** You will be prompted with some on-screen instructions for full set up of the alternate Face ID feature.

   o For people that have one or more physical limitations – especially blind-ness or someone with poor sight - simply click on **Accessibility options** when you are setting up Face ID. From the Accessibility options, you will be able to prevent your iPhone 12 from requesting that your two eyes must be opened before it unlocks. To do this, scroll to **Settings,** click on **Accessibility** and toggle off the "**Require Attention for Face ID**" switch.

How to temporarily disable Face ID

Use the steps below to disable your Face ID temporarily and disable the Face ID from unlocking your iPhone;

1. Press (with your fingers) and hold on the side button plus either of the volume buttons (either the volume up or volume down button) for about three seconds.

2. Once you see the slider, lock your iPhone immediately by pressing on the side button.

   Your iPhone will lock automatically when you do not touch your screen for one minute or thereabout. The Face ID will be re-activated whenever you use your passcode to unlock your iPhone.

How to Turn off Face ID

1. Go to **Settings** and click on **Face ID & Passcode**.

2. Do any of the options below;

   - *Turn off Face ID for specific items only:* This will disable the **Face ID** for any feature or app you choose under this option. You can turn off Face ID for iPhone Unlock, Safari AutoFill, iTunes & App Store and Apple Pay.

   - *Turn off Face ID:* Tap on Reset Face ID.

Note: You won't be able to use the Touch ID & Passcode option for your iPhone 12.

Controlling access to information when your device is locked

Even when your device is locked, there are some features that you can access easily from the lock screen (features such as widget, control center and media control).

Due to security reasons, you will not have access to accept and control USB connection when your device's screen is locked. You will have to unlock your device if you want to accept or reject a USB connection request. Also, you will be able to control what information is accessible from the lock screen and to what extent or how much of the information is visible on the lock screen.

Turning off lock screen access for a particular app or feature on your iPhone will usually disable you or anyone else who might want to gain control of your device from accessing the full detail from any app that brings notification on the lock screen. For instance, when you turn off lock screen access for the calendar app, you won't be able to see full information from the calendar app on the lock screen anytime it prompts notification for you on the lock screen. To control access to information that will be accessible on the lock screen, follow the steps below;

- Go to **Settings** , click on "**Face ID & Passcode**" and then select which feature you want to turn on turn off access for.

You will be able to turn on/off access to the following menus/features when your device is locked;

- **Widgets:**

Turning off access to the widget on the home screen will prevent you from getting full information for the widget on your home screen. You can check below for information on how you can add widgets to your home screen;

Add widgets to the iPhone Home Screen

A particular widget, called **Today view,** gives you updated information from your favorite apps on the home screen at a go. Information ranging from weather, today's headline, upcoming calendar events and a lot more are accessible from the widget on your home screen.

Launch Today View

You can bring **Today View** when you swipe right from the left edge of your iPhone's Home Screen or the left edge of the Lock Screen.

How to move a widget from Today View to the Home Screen

1. Launch Today View by swiping right from the left edge of your iPhone's Home Screen or the left edge of the Lock Screen, then

scroll up or you can search to find the widget you want to move.

2. Position your finger appropriately on the widget and hold till that particular widget starts to shake, and then slide the widget to the right edge of your screen.

3. Place the widget anywhere you want it to be on your home screen and then click "**Done**" to finish.

Tip: The widget called Smart Stack (widget with dots next to it) is a widget that utilizes information such as your location, time and activity to bring the most relevant widget at appropriate periods and times of your day. You can add a Smart Stack to your Home Screen, then swipe through the Smart Stack to check the widgets it contains.

Add a widget to a Home Screen page

1. Go to the page on your Home Screen where you want the widget to be added, and then place your finger on the background of the home screen until all of your apps start to jiggle.

2. Tap on the positive symbol + located at the top of your iPhone's screen to access the widget gallery.

3. Scroll or search to get the widget you want, select the widget by clicking on it, and then swipe left through the size options. Each size you see shows different information.

When you see any size that you want to pick, click on "**Add Widget**," and then tap "**Done.**"

How to remove a widget from the Home Screen

1. On your iPhone Home Screen, bring a quick access menu by tapping and touching a widget.

2. Click on "**Remove Widget**" (or Remove Stack if you want to remove a stack widget) and then select "**Remove.**"

Notification Center

Disabling the Notification center on your home screen will prevent you from seeing the full information for the Notification center on the home screen. You can check below for details of what you will be able to do with the Notification center;

Access and respond to your notifications on iPhone

Customize your notifications to keep track of activities and what is new. Notifications allow you to check missed calls, keep track of important dates and events, see important messages and many other functions. Notification settings can also be customized to access only what is important, see important notifications on your device's lock screen and respond to them.

Access all your notifications in a single place

Your device's notifications will always be delivered to you as they come, but any notification you are yet to check will be saved in the Notification center, which can be accessed anytime later.

You can check all your notifications in the Notification center by doing any of these;

- *On your iPhone's Lock Screen: Swipe up from the middle of the screen to access notifications.*

- *On other screens: Simply* swipe down from the top area of your screen. See older notifications by scrolling up on your screen.

Close the Notification center by swiping up from the bottom of your home screen or anywhere on your screen with one finger.

Respond to notifications

Multiple notifications on your lock screen or the notification center will usually be grouped by app to enable easy management and access. Some apps features can also be organized to group various app notifications. You will always see grouped notifications as small stacks with your most recent notification coming on top of others.

You can carry out any of the actions below;

- Click on a group if you plan to expand the notifications from that particular group. This will allow you to see the notifications individually. You can close the group by tapping on **show less.**

- Touch and then hold a notification to see full details and carry out some quick actions if such app supports them.

- Tap on a notification to launch the app where it came from.

How to dismiss, clear, and manage notifications

Perform any of the below actions;

- *Attend to notifications you get when you are using another app:* Pull such a notification from the top of the notification panel to see the full details. Swipe the notification up to dismiss it.

- *Clear notifications:* Swipe the notification or groups of notifications to the left, and then choose **Clear** or **Clear all.**

- *Send your notifications straight to Notification Center:* Swipe the notification or group of notifications to the left, click on **manage** and select **"Deliver quietly."** Doing this will restrict notifications from such apps or groups from showing on your lock screen, lighting up your screen or making some sounds.

  To view and listen to these notifications again, you can swipe left on any notification in Notification Center, choose **"Manage,"** and then choose **"Deliver prominently."**

- *Turn notifications off for an app or notification group: Simply* swipe the notification or group of notifications to the left, tap on **"Manage"** and then select **"Turn off."**

- *Change how an app brings notifications:* Swipe left on any notification, click on Manage, select Settings, and then pick an option. You can select whether you want to enable notifications from the app, where you want the notifications to appear (in Notification Center, for instance), whether you want the notification to be accompanied by a sound etc.

- *Clear all notifications in Notification Center:* Tap on ⊗ , and then select "**Clear.**"

Siri will make suggestions to turn off notifications for any app that you have not used in a while.

Using the App Library to find your apps

From your Home Screen, bring the App library by swiping left until you see the App Library. Your apps are automatically sorted into categories. For instance, you might see your social media apps under the Social category. The apps that you use most frequently will automatically be reordered based on frequency of usage. When you install new apps, they will be added to your

App Library, but you can change where new apps get downloaded into.

Searching for an app in the App Library

- Scroll to the App Library.
- Tap the search field, and then enter the name of the app that you are looking for.
- Click on the app to launch the app.

Delete an app from the App Library

- Navigate to your iPhone's App Library and click on the search field to open the list.
- Touch and then hold on the app icon, then click on **"Delete App."**
- Click on **"Delete"** the second time to confirm.

Use and customize the Control Center on your device

Get quick access to most used apps and controls – such as airplane mode, flashlight, do not disturb, volume and the screen brightness feature.

How to launch the Control Center

To launch the Control Center, simply swipe down on your screen from the top-right edge. You can dismiss the control center by swiping up from the bottom of the screen.

Access more controls in Control Center

Each of the controls in the control center - such as the airplane mode, volume, flashlight, screen brightness and the "Do not disturb" feature – has additional options to let you utilize and customize each control. For instance, in control center, you can;

- Open any control from the top left group controls by touching and holding such control. For instance, you can launch the

AirDrop options by clicking on the AirDrop icon at the top left group of controls.

- Record a video; take a selfie or a picture by touching and holding on the **camera** icon at the bottom of the control panel.

How to add and organize controls

Customize your control center by setting more controls and shortcuts to apps like Calculator, Voice Memos, Notes and lots more. To achieve this;

1. Scroll to **Settings** and tap on "**Control Center**."

2. Add or delete any control by clicking on the or beside a control.

3. To rearrange a control, tap on the and then drag the control to a new position.

How to temporarily disconnect from a Wi-Fi network

- When you open the Control Center, choose 🛜; to reconnect, click on it again.
- To see the name of the Wi-Fi over which you are connected, simply tap on the Wi-Fi icon.

Disconnecting Wi-Fi from a network will not turn off the Wi-Fi as the Wi-Fi can still join any other available network when you change your location or restart your device. To turn off your Wi-Fi;

- Scroll to **Settings** ⚙ and tap on Wi-Fi. You can as well turn off your Wi-Fi from the control center by tapping on 🚫. While travelling on an airplane, you might be instructed to turn off your Wi-Fi and put your device on airplane mode.

How to temporarily disconnect from Bluetooth devices

When you open the Control Center on your iPhone, tap on the Bluetooth icon. Connect by tapping on the Bluetooth button once again.

Mere disconnecting your iPhone from connected devices does not disable Bluetooth. To disconnect Bluetooth completely, you will need to turn the Bluetooth off. To turn off Bluetooth;

- Go to **Settings** **app** on your device, tap on **Bluetooth** and then turn off the Bluetooth.
- To turn on the Bluetooth again, tap on in the Control Center.

Turn off access to the Control Center in apps

Scroll to **Settings**, select "Control Center," and then turn off Access within Apps.

Siri

Siri is an artificial intelligence system on iOS and iPadOS devices that allows users to carry out and access most things on their gadget without necessarily lifting a finger or tapping on their devices. Siri is not only available on the latest iOS and iPadOS series but also available on old devices. With Siri, you can run your home, automate your cars, and offices by just a text command and voice.

Let us examine some basic Siri commands;

1. Siri calls and texts

You can deploy Siri to make calls and send some important messages.

- Using Siri to call: just say "Hey Siri, call Mercy on speaker" Siri will call Mercy for you and put it on speaker.
- Use Siri to remind yourself to send some important calls: Siri scans for important calls that you have scheduled in your calendar and will notify you prior time.
- Use Siri to text: just say "Hey Siri, text John that I am coming by 8 PM" or you say "Hey Siri, message Dad I will be home in 20 minutes on WhatsApp," or you say "Hey Siri, read the last message from David."
- You can also use Siri for Facetime: say, "Hey Siri, Facetime Richard by 2PM".

2. Siri alarm, reminder, and timers

- Say, "Hey Siri, wake me up by 5PM."
- Request Siri to remind you about a meeting. Say, "Hey Siri, remind me I have a meeting with the panel by 6PM."
- Siri setting timer for you. Just say, "Hey Siri, set a timer for 3 hour."

3. **Siri music:** Siri finds your best songs for you, and can even explore your favorite collections of music.

   - Say, "Hey Siri, play me a song from my favorites."
   - Asking Siri what year a song was composed. Say, "Hey Siri, when did this song come out."
   - Asking Siri what song is this. Just say "Hey Siri, what song is this"
   - Asking Siri to play an artist's latest song. Just say "Hey Siri, play me Justine's latest song."

How to Set up Siri

Most people would have set up Siri during the initial setting up of their device. However, if you have not set up Siri, you can do it now by following the steps below;

- Navigate to **Settings** app on your device, tap on "**Siri & Search**," and then carry out any of the following:

- *If you want to be able to call Siri with your voice:* Turn on the Listen for "Hey Siri."

- *If you want to be able to call Siri with a button:* Turn on the "Press Side Button for Siri" (since your iPhone 12 is with a face ID). For some models of iPhone, they will need to select the "Press home button for Siri" if the model has a home button.

Call Siri with your voice

Calling Siri with your voice will prompt Siri to give out a loud response.

1. Just say "Hey Siri," then prompt Siri by asking a question or request Siri to carry out a certain task for you.
   For instance, you can say something like, "Hey Siri, wake me up by 2PM." Or Prompt Siri to remind you about a meeting. Say, "Hey Siri, remind me I have a meeting with David by 9aM."

2. If you want to ask Siri another question or to tell Siri to do some other tasks for you, voice "Hey Siri" again or click on .

Note: If you plan to restrict your iPhone from responding to your "Hey Siri," place your iPhone face down, or scroll to **Settings**, tap **Siri & Search**, then turn off Listen for "Hey Siri."

Calling Siri with a button

When you call Siri with a button, Siri gives out some loud response when your device is in ring mode and then gives out a silent response if your device is in silent mode.

1. Use the Side button on your iPhone 12 to prompt Siri to action. Press the side button and hold the side button.

   When Siri shows up, ask Siri any question or prompt Siri to do any task for you.

   For instance, you can say something like "What is 8 percent of 600?" or "Set the timer for 10 minutes."

   If you want to ask Siri another question or to tell Siri to do some other tasks for you, say "Hey Siri" again or click on .

Make a correction if Siri misunderstands you

- *Rephrase your prompts:* Click on the , and then say what you want Siri to do for you clearly and in a good tone.

- *Spell out part of your prompts:* Click on the , and then say your question again by spelling out any ambiguous words that Siri didn't pick initially. For instance, say "Call," then spell out the name of the person.

- *Change a message before you send it:* Simply say "Change it."

- *Edit your prompts or questions with text:* If you see your prompts on screen, you can actually edit them. Click on the request, and then use the onscreen keyboard.

Type instead of speaking to Siri

1. Scroll to **Settings**, tap on "**Accessibility**," select "**Siri**," and then turn on the "**Type to Siri**."

To ask Siri any question or request Siri to do any task, use your keyboard and write the text inside the text field.

Replying to messages

How to send and receive text messages on iPhone

The Messages App on your device can be used to send and receive text messages, photos, videos and audio messages. You can personalize each message you send by using animated effect, iMessage apps, Memoji sticker etc.

How to send a message on your device

Use the guide below to send a message to an individual or group of people;

1. Tap the **start a conversation** icon located at the top section of the message screen to start a new message, or tap on an existing message.

2. Fill in the phone number, Apple ID or contact name of the recipient (for individual conversation) or each of the recipients (for group conversation). Or, tap on the **add** icon, then select contacts by tapping on them one after the other.

3. Tap on the text field, write your message and then tap on to send.

   - You will receive an alert as a reply if your message cannot be sent. Simply tap on the alert to send the message the second time. You get an alert like this most times if you do not have enough cellular charges on your phone to send the message. Some messages require you to have a minimum voucher subscription on your device.

Tip: You will be able to see the exact time a message was sent or received when you drag the message bubble to the left of the message screen.

To view the details of any conversation, tap on the name or the phone number of the sender located at the top side of the screen and then tap . You can click on the contact if you wish to edit the contact card, view attachments that accompany the message, share your location, leave a group conversation and lots more.

You will be able to return to the Messages list from any conversation by tapping on the back icon or swipe from the left edge of the message screen.

Reply to a message with Siri

You can ask Siri to reply to a message on your behalf. This is very easy and doesn't take much time. To ask Siri, you can try and say something like;

- "Send David a message telling him to meet in the office by 9AM."

- "Read my last message from John"

- "Reply that is a good song"

Or carry out the following actions:

1. From the list of messages on your message app, click on the conversation that you wish to reply to.

   You can search for contents and contacts in your conversations by pulling down the Messages list and enter whatever it is that you are looking for in the search field. Or, choose from the suggested contacts, photos, links, and others.

2. Tap on the text field and then input the message you want to write.

Tip: You can replace text with emoji by tapping on 😃 or 🌐 and then click on each highlighted word.

3. Tap on the **upload** icon ⬆ to send the message.

You can reply to a specific message by using a Tapback expression (for instance, a heart or a thumb up). Tap twice on the message bubble that you want to respond to and then choose a Tapback.

Share your name and photo

In the Messages app, you will be able to share your photo and name when you initiate or reply to a new message. The photo can be a customized image or a Memoji. When you open the Messages app as a first time user, use the instructions on your device as a guide to choose your photo and name.

You can change your name, photo or sharing options by opening Messages, tap , choose "**Edit Name and Photo**," and then perform any of the listed;

- *Change your profile image*: Choose "Edit" and you will be shown some options to choose from.

- *Change your name*: Tap on the text fields where you see your name.

- *Turn sharing on or off*: Tap on the button beside the "**Name and Photo Sharing**" switch. When the feature is already turned on, you will see a green switch to this effect.

- *Change who can see your profile*: Choose an option below the "Share Automatically" menu. The "**Name and Sharing**" option must be enabled before you can be able to change who can see your profile.

Your message name and photo can be used for your Apple ID and My Card in contact.

How to switch from a Message conversation to a FaceTime or audio call on your device

While you are in a Message conversation, you can initiate an audio call or a FaceTime call with the person you are chatting with.

1. When you are in a Message conversation, tap on the profile picture or the name

located at the top section of the conversation.

2. Choose FaceTime or audio.

Home Control

Introduction to Home on iPhone

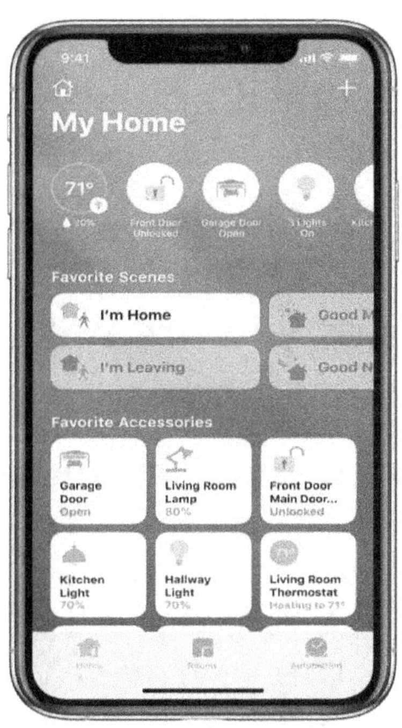

Automate HomeKit - Enabled accessories, like smart plugs, smart TVs, Cameras, locks, thermostats, lights etc by using the Home app to give you a secure means of controlling these accessories all in a single click. You will also be able to group multiple speakers to start playing the same audio, receive a notification when a doorbell camera supported by the Home App identifies a visitor at your doorstep, access and record videos from a supported security camera. With the Home feature, you can use the Apple HomeKit accessory to control any assignment on your iPhone.

You can leverage on the Apple Home app to control Apple-enabled Home tools in your house even when you are not at home. With your Home app, you can control lighting, locks to your room or house, Apple TV, cameras, etc.

Once you have successfully organized your home app with the features you want to be controlling in your house, you can begin to control your accessories individually, or control multiple features at a go with a single command. For instance, you can set the home app to use a command as simple as "Move up" to raise the temperature for your kitchen oven, lock your door, open thermostat, and play your favorite music.

To have total control of your home when you are not around, you must have either an Apple TV, iphone or another iPad that you put at a designated place in your house. You can schedule events to happen simultaneously at a specific time when you are not at home, or even authorized trusted friends or family members to take control in your absence.

Adding an accessory to Home

Let us see how you can set up accessories with the Home app on your iPhone 12;

Any of the accessories you plan to add to your Home app must be an accessory that is capable of using your Wi-Fi network, has a power source, and can be turned on and off at will.

- Select the Home tab 🏠, and then tap ⊕.

- Click on "Add Accessory," and then proceed with the instructions displayed on the screen.

- Any added accessory to Home is assigned automatically to a default room on your Home App or any room you selected.

- You will be requested to use your device's camera to scan the QR code or to input the 8-digit setup code embedded on the accessory. To check the accessory's QR code, check the box or the manual that came with the accessory. You will then be required to join the accessory to a room in your iPhone's Home app. You can give the room a specific name that you will be using to control the accessory with "Hello Siri."

To control an accessory from the Home app, follow these tips;

Tap on the Home⌂ or Rooms display, and then select the button for the accessory you want to control—your thermostat, maybe—to quickly turn on or turn off the accessory, or you can touch and hold on to the button until the control comes up. With this control, you will be able to raise or drop the temperature level of your thermostat.

Edit home accessories

You will be able to edit accessory settings by selecting and holding on the accessory's button, swipe the accessory up on the screen or select ⚙, then perform any of the following activities:

- *To rename an accessory:* Tap on the remove icon ✕ to delete the old name, and then enter another name.

- *To Change an accessory's icon:* Tap on the icon beside the name of the accessory you want to change the icon for. If you cannot see a list of icons you can select from, it means you cannot change the icon for that particular accessory.

Organize rooms into zones

Control different sections of your home with Siri by grouping rooms together into a zone. For instance, if you live in a two-storey home, you may assign the rooms on the first floor to a downstairs zone. Then you can prompt Siri to "Turn off the lights downstairs." To do this,

1. Tap on home icon .

2. Choose a room, click on and then choose Room Settings.

3. Select Zone, then select an existing zone, or click on "Create New" to add that room to a new zone.

Edit a room

You will be able to modify the name for a room and its wallpaper, remove the room or add the room to a zone. When you delete a room, the accessories assigned to the room will move to the Default Room.

1. Click on 🏠.

2. Tap a room, choose 🏠 and then click on "Room Settings."

Group accessories

Group multiple home accessories so that you will be able to control them with a single tap.

1. Touch a home accessory, hold on to the accessory, swipe up on your screen or click on the **settings** ⚙ icon and select **"Group with other accessories."**

2. Select the accessory you wish to group with this accessory— it can be another light in the bathroom or kitchen.

3. Fill in an appropriate name for the group in the Group Name field.

4. Select "**Done**." Toggle on the "**Include in Favorites**" if you want this group to be displayed in your Home tab.

View your home status

The Home app will usually bring some issues that might require your immediate attention – say for instance, you leave your door unlocked, one of the accessory's batteries is draining, or one of the lights outside is on during daytime when it is not needed. You will be able to see all of these issues from the home status where you can attend to them with a single tap.

1. From the Home app ⌂, tap on the Home tab.

2. Tap on any of the round buttons just below the name of your home. You can press and hold a button that represent a group of accessories—three lights that are switched on in two rooms, for instance—and you will be able to control the accessories for each room separately. In this instance, you can turn off the two lights in the living room but leave the one in the kitchen. You can control all of the accessories in the group by tapping a

single button. This button will be able to turn on all the lights or turn them off.

Set up security cameras in Home on iPhone

The Home app on your iPhone 12 can be used to access video activities recorded from the security cameras in your house when it detects the movement of persons, vehicle or animals. The video taken by your cameras will be analyzed privately and encrypted on your home hub device (Apple TV, HomePod, or iPad). The videos will be uploaded on your iCloud for security purposes; to allow only you and those you share it with to see it.

Camera options

Once you have successfully added a compatible camera to your Home App, you will be requested to assign streaming and recording options and then assign the camera to a room. The camera, by default, will be marked as favorite and will be shown in your Home tab. You can edit the settings by touching and holding the camera button, and choose ⚙ to access the options below:

- *Room: You can locate a* camera in a room inside your home, or create a room for a location outside, like your backyard.

- *Notifications:* Select "Notifications" to indicate when you will like to receive a notification (you can choose to be receiving notifications at any time when you are absent from home) when a motion is detected, or when a video clip is recorded.

- *Streaming & Recording: You can choose between Off, Stream, Detect Activity* and Stream & Allow Recording.

Different settings can be created for when you are available at home and when you are absent at home. For instance, you can decide to pause streaming and recording when you are at home, but continue the steaming from a camera outdoors.

Note: The Home app deploys the actual location of devices that are members of the home to switch between Home and Away settings. For instance, when you leave your house for your work with your iPhone 12, your camera moves from the When Home settings to the When Away settings.

- *Recording Options: When motion is detected, your camera will be able to record it.* When you select Specific Motion, the motion of persons, vehicles or animals will usually trigger video recording.

Create activity zones

Create zones that allow your camera to concentrate only on the most important sections within its view – for example, your camera can be made to focus only on the walkway in the front and not the sidewalk. You will be notified when any motion is noticed in these zones.

Note: Activity zones can only be created only for those cameras that have been set to record when they detect some specific motions. Motion detection automation won't be affected by Activity zones. To create Activity zones, follow the steps below;

1. Click on the Home tab, touch and hold the camera and then choose ⚙.

2. Select "Activity Zones," tap or draw on your video to create a zone and then click on "Add Zone." You can click on "Invert Zone" if you want your camera to notice motions only outside of the zone you selected.

3. You can create some additional zones - if you want - within the camera's view—one for the driveway, maybe and another zone for the mailbox.

4. Click on "Add Zone" for each of the zones you have created and then tap "Done."

View video

1. Tap on the Home tab and then click on the camera. Live video will play automatically.

2. Browse recorded clips by swiping through the timeline at the lower end of your screen.

3. Play a video clip by tapping on it.

4. Select Live to go back to live video.

Choose access options

To let other people view video from your cameras, use the following steps;

1. Tap on the Home App icon.

2. Click on Home Settings, and tap on a person below the "People" heading.

3. Tap on "**Cameras**," and then select an option.

Turn ON Adaptive Lighting

Some lights will usually allow you to adjust their color temperature, when you are able to change from one color to another. Adjust the color temperature throughout the day by adjusting settings for lights that support adaptive lightning. A popular Smart lighting bulb is the IKEA smart lighting bulb, which can usually work with the Adaptive lighting feature on iPhone. With Adaptive lighting, you will be able to turn on lights in your house or room to help you with the normal morning wakeup activities, switch to a cooler color temperature in the day, and help yourself to sleep by switching to a warmer color tone in the night. Apple only allows Adaptive lighting features for Smart lights and not for daylight-only bulbs.

Note: The Adaptive lighting feature is still currently being advertised by Apple as at the time of writing this book.

1. Touch any lightning accessory that supports Adaptive lightning by placing your finger on it and then hold.

2. Click on .

Add more homes with iPhone

You can actually add many physical spaces – say a home and a small office space- in the home app .

1. Click on the home icon and then select Home Settings.

2. Select "Add Home," input an appropriate name for the home and then tap on Save.

3. If you wish to switch to another home, you can simply tap on and then choose the home you want.

CHAPTER THREE

BASIC iPhone 12 SETTINGS

Adjust the volume on iPhone

The buttons on the side of your iPhone can be used when you want to adjust audio volume when you are on calls, watching videos, or listening to music. You can also request Siri to adjust the volume for you. To request Siri to adjust the volume level, you can say, "Hey Siri, raise the volume" or say "Hey Siri, reduce the volume" You can as well navigate to the control center to adjust volume, or you can even customize your Control Center by adding the volume up/down function. You can also change the volume by going to Settings and then to **music > volume limit**, and you can then use the slider to set the maximum and minimum volume.

Lock the ringer and alert volumes in Settings

1. Navigate to the **Settings** **app** on your device.

2. Tap on **Sounds & Haptics.**

3. Tap to disable the "**Change with Buttons**" option.

How to adjust the volume in Control Center

The Control Center can be used to adjust device volume especially when your phone is locked or when you are currently using an app. To adjust the volume in the Control Center, simply launch the control center and then drag the speaker icon.

Limit the headphone volume

The headphone volume can be reduced for videos and music.

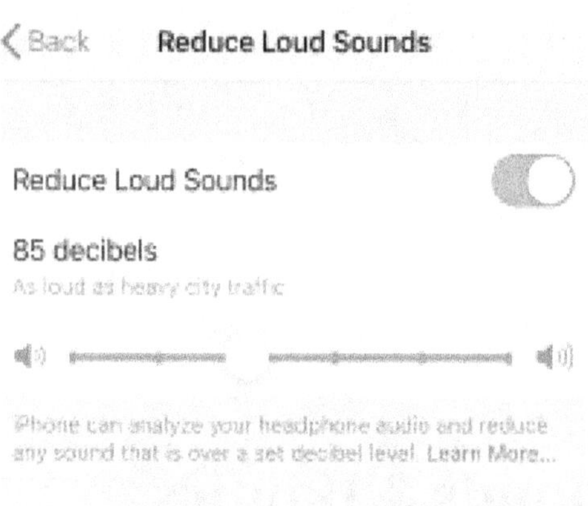

1. Navigate to the **Settings** app on your device.

2. Select **Sounds & Haptics.**

3. Choose "**Reduce Loud Sounds**," turn on the "**Reduce Loud Sounds**," and use the slider to choose the maximum decibel level for your headphone audio.

Change iPhone sounds and vibrations

Your iPhone 12 will always make some sounds whenever some notifications in the form of calls, messages, email, voicemail, reminder etc enter your device. This sound can be changed if you want by making some adjustments in the settings.

A sense vibration – called **haptic feedback -** is usually noticed whenever you do some touch activities on your phone. Try and touch the camera icon on the home screen and you will most often sense the haptic feedback.

Set sound and vibration options

1. Scroll to **Settings** , click on "**Sounds & Haptics.**"

2. Set the volume limit for all sounds by dragging the slider below the "**Ringers and Alerts.**"

3. Select a sound type – maybe a ringtone or text – if you wish to set the vibration patterns and tones for sounds.

4. Perform any of the listed actions below;

   - Select a tone (scroll up and down on the screen to see all of the available tones). Ringtones will be played for clock alarms, incoming calls and the clock timer while text tones are usually deployed for text messages, new voice-mail and some other alerts.

- Tap on "**Vibration**," and then choose a vibration pattern, or tap on "**Create New Vibration**" to customize your own vibration.

Turn haptic feedback OFF or ON

1. Navigate to **Settings** and tap on "**Sounds & Haptics.**"

2. Turn on/off System Haptics. By disabling the System Haptics, you will no longer hear or feel vibrations for all of your incoming calls and alerts.

Tip: In case you can no longer access or hear incoming alerts and calls just like you used to do, it is possible that you have mistakenly enabled the "Do not disturb" feature. Launch the Control Center and check whether the "**Do not disturb**" is enabled. If is highlighted, click on it to **turn off** the "**Do Not Disturb**." You will also see the in the status bar when the "Do not disturb" is active.

How to adjust screen brightness and color on iPhone

On your iPhone 12, you can dim your screen to improve and prolong the battery life, use Night Shift, set Dark Mode and adjust the screen automatically for your lighting conditions.

Turn Dark Mode ON or OFF

The Dark Mode feature gives you some dark color experience that is perfect for low-light environments. You will be able to turn on Dark Mode from the Control Center or set the Dark Mode to automatically turn on at night (or on a particular time you customized) in the Settings app. When you allow Dark Mode, you will be able to utilize your iPhone at night (maybe while you are on bed) without the light interfering with other peoples' activities.

Perform any of these actions;

- Launch the Control Center, touch and hold

, and then tap ◑ if you decide to turn on/off the Dark Mode.

- Scroll to **Settings** ⚙, tap on "**Display & Brightness**," and then select "**Dark**" to turn on the Dark Mode, or click on **Light** to turn off the Dark Mode.

Schedule Dark Mode to turn on and off automatically

1. From the **Settings** ⚙ app on your iPhone, click on "**Display & Brightness**."

2. Turn on "**Automatic**," and then tap on "**Options**."

3. Choose between "**Sunset to Sunrise**" or "**Custom Schedule**."

   If you choose the Custom Schedule option, tap on the options to schedule the exact times that you want the Dark Mode to turn on/off. If you select Sunset to Sunrise, your device will be able to use the data from your clock and your geographical location to know when it is night in your area.

How to adjust the screen brightness manually

If you wish to make your device screen brighter or dimmer, do one of the following settings;

- Launch your Control Center and then drag ☀.

- Scroll to the **Settings** ⚙ app on your device, tap on **"Display & Brightness,"** and then drag the brightness slider.

How to automatically adjust your device's screen brightness

The built-in ambient sensor will be used by your phone to adjust your screen brightness for your current light conditions.

1. Navigate to **Settings** and choose "**Accessibility**."

2. Tap on "**Display & Text Size**," and then turn on "**Auto-Brightness.**"

Turn True Tone ON or OFF

On your iPhone 12, you will be able to automatically adapt the intensity and color of the screen's display to match the environment's light by turning on True Tone.

Carry out any of the actions below;

- Launch the Control Center, touch and hold , and then select to turn on/off True Tone.

- Scroll to **Settings** , tap on "**Display & Brightness,**" and then turn on/off True Tone.

Turn Night Shift on or off

The Night shift can be turned on manually to assist you in a dark room during the day time.

Launch the "**Control Center**," touch and hold ☀, and then click on 🌙.

Schedule Night Shift to turn ON and OFF automatically

The color of your phone's display can be adjusted to the warmer side of the spectrum so that you will be able to view your phone's screen without having a negative effect on your eyes.

1. Navigate to **Settings** ⚙, choose "**Display & Brightness**" and tap on "**Night Shift**."

2. Turn on "**Scheduled**."

3. Simply drag the slider located below Color Temperature to the cooler or warmer side of the spectrum to adjust the Night's shift color balance.

4. Select "**From**," then select between "**Sunset to Sunrise**" or "**Custom Schedule**."

If you chose the Custom Schedule option, tap on the options to schedule the exact times that you want the Night Shift to turn on/off. If you select Sunset to Sunrise, your device will be able to use the data from your clock and your geographical location to know when it is night in your area.

Note: You won't be able to access the "**Sunset to Sunrise**" option if you have disabled Location Services in Settings > Privacy, or if you have turned off Setting Time Zone in Settings > Privacy > Location Services > System Services.

How to change the name of your device

When you change your device's name, the new name will be used by iCloud, Hotspot, Bluetooth and your computer. The name of your device can be changed by following the steps below;

1. Scroll to **Settings**, click on "**General**, tap on "**About**" and click on "**Name**."

2. Tap , fill in a new name for your phone and then click on "**Done.**"

Set the date and time on iPhone

Although your phone will automatically set the date and time you see on the lock screen by default based on your active location. But it is possible that something goes wrong with the date and time and you will have to adjust it. To adjust date and time for your phone, follow the steps below;

1. Go to **Settings**, click on "**General**" and then tap on "**Date & Time**."

2. Turn on any of the options below;

   - *Set Automatically:* Your device will update the actual time through your active network and update the time for the time zone in your region. Some networks do not really support network time, so your device might be unable to determine your local time automatically in some countries or regions

- *24-Hour Time:* (this might not be available in all countries or regions) your iPhone will display the hours from 0hr to 23hr.

You will be able to change your default date and time by turning off the "**Set Automatically**," and then change the date and time.

Set the language and region on iPhone

When you first purchase your new device, you will be requested to choose your preferred language and region. But when you move out of your current location, you will need to update your device and change your language or region. To do this, simply follow the steps below;

1. Go to the **Settings** app on your device, tap "**General** and then click on "**Language & Region**."

2. The following settings can be edited;

   - Your device's language.

   - Your new location or region

   - The calendar format

   - The temperature unit (Celsius or Fahrenheit)

If you wish to add another language with a new keyboard to your device, select "**Add Language**," and then choose a language.

How to set up mail, contacts, and calendar accounts

There are some apps that you will meet by default on your device, and in addition to these apps; your device is equally compatible with Microsoft Exchange and some of the most famous internet – based mail, calendar service and contacts. You will be able to set up accounts for each of these services on your device.

How to set up a mail account

1. Scroll to the **Settings** app on your device, tap on "**Mail**" click on "**Accounts**" and choose "**Add Account**."

2. Do any of the actions below;

   - Select an email service—for instance, iCloud or Microsoft Exchange—and then fill in your email account information.

   - Choose "**Other**," tap on "**Add Mail Account**," and then fill in your information to set up your new account.

How to set up a contacts account

1. Select the **Settings** app from your phone's home screen, click on "**Contacts**," select "**Accounts**" > "**Add Account**" and then click on "Other."

2. Tap on "**Add LDAP Account**" or choose "**Add CardDAV Account**" (if it is supported by your organization) and then fill in your server and account information.

How to set up a calendar account

1. Navigate to **Settings**, click on "**Calendar**", tap on "**Accounts**" and select "**Add Account.**"

2. Click on "**Other**," and then carry out any of the following:

   - *Add a calendar account:* Tap on Add CalDAV Account and then input your server and account information.

   - *Subscribe to iCal (.ics) calendars:* Tap on "Add Subscribed Calendar," and then fill in the URL of the .ics file you want to subscribe to; or you can import an .ics file from Mail.

Access features from the iPhone Lock Screen

The iPhone Lock Screen, which allows you access to the current time, date and important notifications from your most recent apps and features, will come up just anytime you wake your device or turn it on. Even when you are in the lock screen, you can still access your camera app, view notifications from some enabled apps, launch the Control Center and get updated information from any app you used frequently at a glance.

How to access features and information from your device's Lock Screen

The iPhone lock screen will always give you access to some common and important features, even when your device is locked. The following actions can be carried out from the lock screen;

- *Open Camera:* Swipe left on the lock screen. Touch and hold the camera icon and then lift your finger. This action will launch the camera app.

- *Open Control Center:* To open the Control Center, simply swipe down from the top right section of your device.

- *View earlier notifications:* Swipe up from the center of your iPhone's screen.

- *Access Today View:* Swipe right on the lock screen.

When you control access to information available on the lock screen, you can always customize the information that will be shown to you on the lock screen. To control access to information that can be accessible from the lock screen, simply navigate to **Settings** **and** click on "Face ID & Passcode."

How to show notification previews on your Lock Screen

1. Go to **Settings** and click on "**Notifications.**"

2. Select "**Show Previews**," and then click on "**Always**."

The Notification previews can include text from Messages, details about calendar invitation and lines from Mail messages.

Open apps on iPhone

In this section, you will be able to familiarize yourself with your Home Screen and apps. The Home Screen displays all of your apps arranged into pages. When you need more spaces for apps, more pages are automatically added on the next screen.

The App Library also provides an easy platform where you can view and access your apps. The App Library is a space located at the end of your iPhone's Home screen pages where your apps are arranged in a concise and easy-to-use view.

How to launch apps on the Home Screen

1. To go to your Home screen on your iPhone 12, kindly swipe from the lower end of the screen.

2. Swipe to the right or left to browse apps on other Home Screen pages.

3. Tap on the app icon to open the app.

4. You can return to your first Home Screen page by swiping up from the bottom edge of your screen.

You will also be able to move your apps, organize them and remove some of them. Check the steps below to see how you can **move and organize your apps** on your device;

How to move apps around your Home Screen, into the Dock, or to other Home Screen pages

1. On your Home screen, place your finger on any app to touch and hold on the app and then select "**Edit Home Screen**." The apps you selected will begin to jiggle.

2. The app can be dragged to any of these locations;

   - A new location on the same page

   - The Dock located at the bottom of your iPhone's screen

   - A new Home Screen page

     Drag the app that you want to move to the right edge of the screen. You might have to wait for some seconds before the new page shows up. You will see some dots just above the Dock, which show how many screen pages you have and which of the pages you are currently viewing. For instance, four dots above the Dock indicate that you have a four pages screen. When you are done, simply tap on **"Done."**

How to create folders and organize your apps

You will be able to group and arrange your apps inside a folder on your home screen so that they can be accessed more quickly. In fact, you can even customize a name for the folder.

1. On your Home screen, touch and hold on an app and then choose **"Edit Home Screen."** The apps will begin to jiggle.

2. Drag an application on the home screen over another application to create a folder for apps.

3. Drag other applications into the folder. Your apps folder can actually contain multiple pages of apps if one page is not enough to take care of all apps inside the folder.

4. Tap on the name field if you want to rename a particular folder and then fill in a new name for the folder.

5. When you are done, click on **"Done."**

If you want to disband/delete a particular folder, simply drag all the apps out of that particular folder and the folder will be deleted automatically. Once you take out the last app, the app folder will be removed automatically.

Note: The app layout in the App Library is not affected when you organize your apps on your home screen.

Explore the App Library

The App Library will automatically arrange your apps by category, such as Social, Entertainment, Tools, creativity etc. The apps that you deployed most often will be displayed near the top of the screen and at the top of their categories. This will allow you to easily find and use them.

To locate your App Library, go to your Home Screen and swipe to the left past all the Home screen pages.

In the App Library, you will be able to carry out any of the following:

- *Launch an app:* Tap on the app to open, if the app is visible.

- *Expand a category:* You can explore all the apps for a particular category that has more apps below the top level (shown by a few small app icons) by clicking on the small icons.

- *Search for apps:* You will be able to search for any app by using the Search bar located at the top section of the screen. Simply fill in the name or initials of the apps you are searching for and you will see the app coming up provided you have such app on your device.

- *Perform quick actions:* Launch a quick action menu by touching and holding any of your apps. Touching and holding a particular app for too long before you select a quick action will make all of the apps to start jiggling or dancing. If you see all the apps jiggling, kindly click on **"Done"** and then try again; but this time you should be careful so that you will not hold an app for long.

- *Add a particular app to your Home Screen:* Launch the quick action menu by touching and holding any of your app, then choose "Add to Home Screen" (you will only be able to see the "**add to home screen**" if that particular app is not already available on the home screen). The app you added to the home screen will still show in the App Library.

- *Delete/Remove a particular app from your device:* Touch the app and then hold it, you will be prompted with some quick menus where you can select "Delete App," and then choose "Delete." Deleting any app will remove the app from the App Library and from the Home Screen. Note that you can as well add new apps from your App Store to both the App Library and the Home Screen or just the App Library alone. To do this; scroll to **Settings** click on "Home Screen," and then select either the "**Add to Home Screen**" or "**App Library only**." You can as well set an app notification badge to show on apps in the App Library by allowing the "**Show in App Library**."

Hide and show Home Screen pages

The App Library gives you a quick means of navigating your applications at a glance, so most users will prefer to hide their Home screen pages to bring their App library closer to the first page of their Home Screen. Too many Home Screen pages imply too many screens to browse when navigating your apps. With the App Library closer to your first page, you can quickly access your apps anytime. To hide Home Screen pages;

1. On your Home screen, touch and hold on an app and then choose "Edit Home Screen." The apps will begin to jiggle.

2. Tap on the dots located at the lower end of your screen. You will get the thumbnail images of your Home Screen pages displaying with checkmarks beneath them.

3. To hide pages, simply tap on the checkmark to un-select the check marks beneath the thumbnails of the pages you don't want. You will be able to show hidden pages by tapping again to add the checkmarks.

4. Click on "**Done**" twice.

By hiding some extra home screen pages from your screen. It becomes very easy to go back and forth from the first page of your Home Screen to the App Library with one or two swipes.

Note: By hiding home screen pages, any new apps you recently downloaded from the

App Store may be added to the App Library instead of the Home Screen.

How to quit and reopen an app on iPhone 12

If you observe that one of your applications is not working as expected, you can decide to quit the app and then open it after quitting. Quitting and reopening the app is one of the common troubleshooting techniques to resolve apps not responding. Quitting an app doesn't necessarily save battery power but it can get an app that is not responding to start responding.

1. If you plan to quit a particular app, launch the app switcher, find the app by swiping to the right and then swipe up on such app. You can access your app switcher by swiping up from the bottom edge of the screen and then take a pause when you get to the center of the screen.

2. You can reopen any app you have already closed by going to your device's home

screen or the App Library and then tap on the app to reopen.

3. If the app still doesn't respond, even when you have quit and reopen many times, kindly restart your iPhone.

How to remove apps from your iPhone

Most of your apps that are accessible from the home screen can be removed from the home screen if you don't need them again. Although, you will still be able to download apps you have deleted from the App store if you want.

How to Remove apps

To remove any app from the Home screen, perform any of the actions below;

- *Remove an app from the Home Scr-een: Bring the quick action menu by* touching and holding the app on the Home Screen, tap on **"Remove app"** and then choose **"Move to App library"** if you want to keep the app in your App Library or tap **"Delete app"** to completely delete the app from your iPhone.

- *Delete an app from your App Library and Home Screen: Bring the quick action menu by* touching and holding the app on your App Library, select "**Delete App**," and then choose "**Delete.**"

- *Add an app in the App Library to the Home Screen: Bring the quick action menu by* touching and holding an app in your App Library, then tap on "**Add to Home Screen**." The **"Add to home screen"** option will only be available if that particular app is not already available on the Home screen.

Deleted apps can be downloaded again anytime you want.

Two categories of apps are enjoyable on your device; Third party apps and in-built apps. Third party apps are apps that you downloaded from the App store. The third party apps are usually uploaded by Apple developers whose apps have been verified on the Apple Store for download. Built-in apps are apps that accompanied your iPhone 12 from the manufacturer. The two categories of apps can be removed from your home screen if you want. The following built-in apps can be deleted from your device; Books, Compass, Calendar, Contact (your contact information will still be available through Phone, Mail, Messages, FaceTime and a few other apps. If you want to remove a contact, you need to restore Contacts), Calculator, Files, Home , Mail, iTunes Store, News, Music, Measure, Notes, Reminders, Podcasts, Stocks, Tips, TV, Voice Memos, Weather, Watch, Shortcuts etc.

How to delete a built-in app from your phone

1. On your iPhone 12, touch the app and hold it to bring some actions.

2. Choose "**Rearrange Apps.**"

3. Click on ⊖ located in the upper-left side to delete the app.

4. Tap "**Delete**" and click on "Done."

If you are using an Apple Watch (probably you have the latest Apple Watch 6), deleting a particular app from your iPhone will also delete the app from the Apple Watch.

Restore a built-in app that you deleted

1. On your iPhone 12, go to the App Store.
2. Search for any app that you want to restore. To ensure proper search, type the correct name of the app in the search field.
3. Choose ⬇ to restore the app.
4. Wait a little while for the app to complete restoration and then open the app from the home screen.

If you have an Apple Watch (probably an Apple Watch 6 or older Apple Watch), restoring a particular app to your iPhone will also restore the app to the Apple Watch.

How to type with the onscreen keyboard on iPhone 12

On all models of iPhone, you can use your onscreen keyboard to input and edit text. The iPhone 12 also supports typing with the magic keyboard and using dictation to write your texts. You will also be able to connect iPhone with external keyboard through Bluetooth.

How to Pair Magic Keyboard to iPhone

1. Ensure that the keyboard is charged first and then turned on. This is because the magic keyboard uses charging to power itself.

2. On your iPhone 12, go to **Settings**, click on "**Bluetooth**," and then turn on the Bluetooth.

3. Select your Magic keyboard once you see it on the device's list.

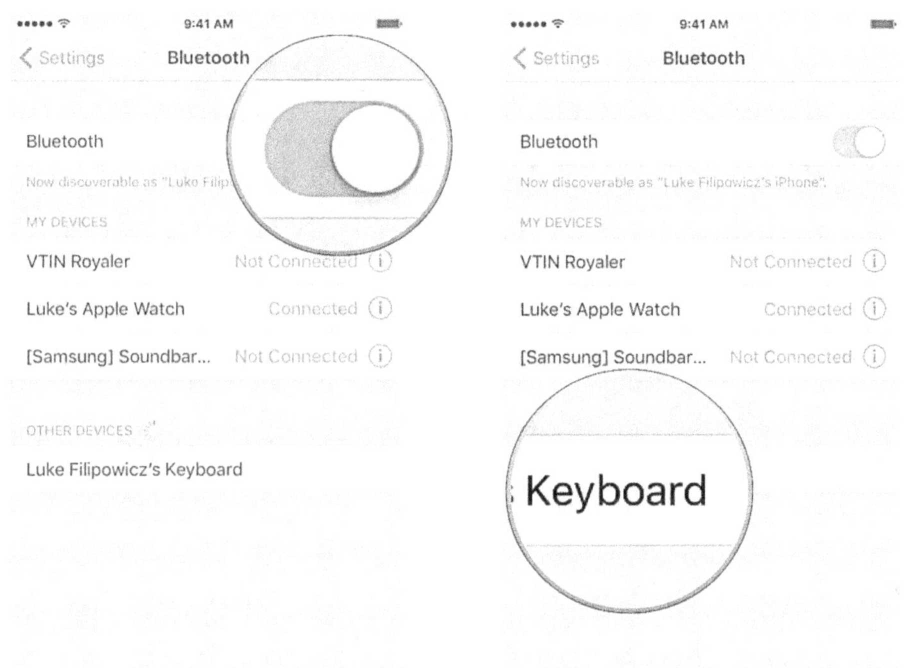

How to enter text using the onscreen keyboard

The onscreen keyboard will be prompted automatically whenever you are using an app that supports text editing. To launch your onscreen keyboard, simply tap on a text field. Tap on each key on the keyboard to write your text or deploy the QuickPath, which allows you to enter a word by simply sliding between letters without necessarily lifting your fingers (this is not supported in all languages). When you are done typing, simply remove your fingers from the keyboard. You can actually use the QuickPath and the "tap a text field" to enter texts and also switch in the middle of texting. Tapping on the ⊗ after sliding to type a word will delete the entire word.

Note: As you are sliding to enter your texts, you can access suggested alternatives to any word you are entering. This will further aid efficient and fast typing.

While entering texts, you can perform any of the action below;

- *Type uppercase letters:* Tap on "Shift," or tap on the Shift key and then slide to a letter.

- *Turn Caps Lock on:* Click twice on the shift to turn on the CAPS LOCK option.

- *Quickly finish a sentence with a space and a period:* Tap twice on the Space bar.

- *Correct spelling:* Tap on a misspelled word (misspelled words are often underlined in red) to see suggested corrections, and then select a suggestion to replace the misspelled word, or type the correction.

- *Enter numbers, symbols or punctuation:* Tap 123 or #+=.

- *Undo your last edit:* You will be able to edit any previous edit by swiping left with your three fingers.

- *Redo your last edit:* Swipe right with your three fingers.

- *Enter emoji:* Click on 😃 or 🌐 to quickly use the emoji keyboard. You will be able to search for an emoji when you input a commonly used word—like "heart" or "smiley face"—inside the search field located above your emoji keyboard. You can then swipe through the prompted emoji.

Take a screenshot or screen recording on iPhone on iPhone 12

While using your screen, you can carry out a screen recording to record your current actions on the screen or take a picture of your screen just as the action is being performed. The recorded video or pictures can be shared later with anyone or sent as an attachment in mails or social media platforms.

Take a screenshot

1. To take a screenshot on your iPhone 12, you can press and release the side and volume up button simultaneously.

   Click on the screenshot from the lower-left corner of the screen, and then select **"Done."**

   Select either Save to Files, Save to Photos, or Delete Screenshot.

If you select Save to Photos, you will be able to access the picture in the Screenshot albums in your Photos app, or you can also see it in the All Photos album if you have turned on the **iCloud Photos**

from **Settings** and then tap on **Photos.**

Tip: You can create a PDF of a page from a website, document, or mail by taking a screenshot, tap on the thumbnail image and then click on **Full Page.**

Create a screen recording

Create a screen recording of activities you are doing on your screen and as well record sound along with it.

1. Navigate to **Settings**, tap on **Control Center** and then click next to Screen Recording.

2. Launch Control Center, select, and then wait for a three-second countdown.

3. To stop the recording, launch the Control

Center, click on ⊙ or tap on the red status bar located at the top part of your screen, then choose Stop.

Navigate to your Photos app on the home screen and then choose your screen recording.

Change or lock the screen orientation on iPhone 12

Your device is able to give a different view on rotation; either from portrait to landscape and vice versa.

Lock or unlock the screen orientation

Locking your device's screen orientation will restrict your phone from changing orientation when you rotate your device. To do this;

Launch the Control Center, and then click on ↻.

Doing more with keyboard on iPhone

12

Turn on Dictation

Instead of typing with the keyboard, dictation provides a way of interacting with your device through texting/writing with your voice. To turn on dictation, follow the steps below;

1. Navigate to **Settings**, tap on "**General**" and click on "**Keyboard.**"

2. Turn on the "**Enable Dictation**" option

Dictate text

1. Choose on your onscreen keyboard and begin to speak. If you cannot see the icon, turn on the **Dictation** option by going to **Settings,** tap on **General** and click on "**Keyboard.**"

2. When you are done, simply tap.

How to create a text replacement

1. When you are typing message inside a text field, simply touch and then hold 😃 or 🌐.

2. Choose "**Keyboard Settings**," and then click on "**Text Replacement**."

3. Tap on ➕ located at the top right.

4. Enter a phrase inside the Phrase field and also the text shortcut you plan to use for it inside the Shortcut field.

How to add or change your keyboards on your device

You will be able to; turn on or off typing features like the spell checking, add more/ extra keyboards to allow you write in multiple languages and also change the layout of the wireless or onscreen keyboard for your iPhone. Adding keyboards for various languages will allow you to type in more than one language without necessarily having to switch between keyboards. The keyboard will adapt itself to switch automatically between two languages you used more often. Some languages are not supported with the multiple keyboards.

How to add or remove a keyboard for another language

1. Navigate to **Settings**, tap on **General** and choose "**Keyboard**."

2. Tap on Keyboards and then carry out any of the following actions:

   - *Add a keyboard:* Choose "**Add New Keyboard**," and then pick a keyboard from the list of keyboards. This step can be repeated to add more keyboards.

   - *Remove a keyboard:* Tap on "**Edit**," click on the ⊖ next to a particular keyboard you plan to remove, choose **"Delete"** and then click on **"Done."**

   - *Reorder the keyboard list:* Select "**Edit**," drag ☰ next to any keyboard that you want to reorder to a new location on the list and then tap on "**Done.**"

When you add a keyboard for a different language, the keyboard is automatically added to your preferred order list. The keyboard's preferred order list can be accessed, and you can add languages for the keyboard by going to **settings,** tap on **"General"** and select **"Language & Region."** You can also reorder the keyboard's list to modify how various apps and websites on your phone bring text.

How to switch to another keyboard

1. When you are writing your message or text, kindly touch and then hold on 😃 or tap 🌐.

2. Tap on the name of the keyboard that you plan to switch to.

You can also click on 😃 or 🌐 to switch from a particular keyboard to another. When you tap continuously, you will be prompted

with other enabled keyboards for your
iPhone 12.

Use AirDrop on iPhone to send items to nearby devices

The AirDrop feature on your device will allow you to send images, locations, website links, videos and lots more to other phones (running iPadOS 13 and above & iOS 7 and advance) or computers that are within range. To use the AirDrop feature, you will be required to enable your Bluetooth and Wi-Fi because the information transfer will be done over Wi-Fi and Bluetooth. You are also required to sign in to your Apple ID before you will be able to use AirDrop. Any transfer done with AirDrop is encrypted for security reasons.

Move text

1. When you are utilizing an app that supports text editing, you can tap on the

text to highlight the areas that you wish to move.

2. Touch and then hold the text you selected until the text lifts up and you can then drag it with your finger to another location within the app.

If you don't wish to move the text again, you can lift your finger before you drag the text or drag such text off your screen.

How to turn off predictive text

1. While editing your text, touch and then hold ☺ or ⊕.

2. Tap on **"Keyboard Settings,"** and then turn off "**Predictive.**"

How to set typing options

You can type while avoiding errors on your device by enabling special typing features, like auto-correction and predictive text.

1. While entering text with your phone's onscreen keyboard, you can touch and

hold 😀 or 🌐 and then choose **"Keyboard Settings."** Alternatively, navigate to **Settings** ⚙, click on **"General"** and select **"Keyboard."**

You will be prompted with the list where you can either turn on the typing features or turn it off.

Using the AirDrop feature

How to send an item using AirDrop

1. Open any images, videos, or just any item at all that you want to send, click on; ⬆ AirDrop, Share, •••, or any other button that depicts the app's sharing options.

2. From the sharing options displayed, click on ◉ and then select the profile picture of the AirDrop user that you want to send the file to. The AirDrop user must not be far from you.

   Tip: Simply point your iPhone 12 in the direction of another similar device (running iOS or iPadOS) and then tap on the profile picture of the iPhone user at the top of your screen. If you cannot see the person as a nearby AirDrop user, request the user to launch **Control Center** on their device and enable AirDrop to receive files.

There are many methods available to send items other than the AirDrop option. If you wish to use another method, simply select the method from the row of the sharing options (where you once selected the Air-Drop option). You can send your items via messages or mail app.
AirDrop can also be utilized to securely share your website passwords and apps with any-one using Mac, iPhone, iPad or iPod touch.

How to allow others to send items to your iPhone 12 using AirDrop

1. Launch the Control Center on your own iPhone 12 and then choose. If you

cannot see ⊚, you can touch and hold the group of controls from the top-left.

2. Select "**Contacts only**" or "**Everyone**" to indicate the identity of the one(s) you want to send the items to.

Change the wallpaper on iPhone 12

On your device, select an image or photo as the wallpaper for your Lock Screen or Home Screen. You can select from still and dynamic images.

Change the wallpaper

1. Scroll to **Settings** ⚙, tap on **Wallpaper** and select "**Choose a New Wallpaper.**"

2. Perform any of the following action:

   - From the top of your screen, select a preset image from a group. You can choose "Still" or "Dynamic. "

The wallpaper shown with ◑ will usually change appearance when you turn on the Dark mode.

- Click on one of your own images (select an album and then select the photo).

  To reposition the image you selected, zoom in on the image by pinch open, and then move the image by dragging it with your finger. Zoom back out on the image by pinch closed.

- Click on ⬀ to enable the Perspective Zoom (some wallpaper choices support this feature). The Perspective Zoom feature will make your wallpaper appear to be moving once you switch your viewing angle.

  Note: If the "Reduce Motion" option is enabled (from **settings > accessibility > motion),** you will not be able to see the Perspective Zoom option. To disable the **"Reduce Motion"** option, go to **settings,** tap on **accessibility,** select

motion and then turn off the "Reduce motion."

2. Click on "Set," and then select one of these;

- Set Lock Screen

- Set Home Screen

- Set Both

If you decide to enable the "Perspective Zoom" for the wallpaper that you have already set, navigate to **Settings**, select "**Wallpaper**," tap the picture of your Lock Screen or Home Screen, then select "**Perspective Zoom**."

Set a Live Photo as wallpaper for the Lock Screen

When you apply a Live Photo as your wallpaper, simply touch and hold your Lock Screen to watch the Live Photo.

1. Navigate to your **Settings app**, tap on **Wallpaper** and tap on "**Choose a New Wallpaper**."

2. Carry out any of these actions;

- Select Live, and then select a Live Photo.

- Click on your Live Photos album and then select a Live Photo (you may have to wait a while before it to download).

2. Select "**Set**," and then click on "Set Lock Screen" or "Set Both."

Multitask with Picture in Picture on iPhone

With the Picture in Picture feature, you can utilize FaceTime or see a video even while you are still using other applications.

When you are seeing a video or utilizing the FaceTime app, tap on .

The video window will usually scale down to one corner of the screen to enable you to access the Home screen and launch some other applications. With the video window displaying, the following actions can be carried out;

- *Resize the video window:* To get a larger window for the small video window, pinch open. Pinch closed to reduce the video window.

- *Show and hide controls:* Click on the video window.

- *Move the video window:* Drag the video window to another corner of your screen.

- *Hide the video window:* Drag the video window off the left edge or the right edge of your screen.

- *Close the video window:* Click on ⊗ .

- *Return to a full FaceTime or video screen:* Click on ◲ in the small video window.

Get apps in the App Store on iPhone

In your App Store app, you can view featured stories, new apps can be discovered and you can also learn new app tricks and tips.

Find apps

Ask Siri. You can say something like: "find Game apps for me on the App store" or say "Get the Cooking app."

You can as well select any of the following:

- *Today:* Find featured apps and stories.

- *Apps:* See new app releases, view the top charts, or browse by category.

- *Search:* Input what you are searching for and then click on Search on your keyboard.

Get more info about an app

The following information and more are usually accessible when you tap on any app;

- Supported languages

- The size of the app

- Previews or screenshots

- App ratings and reviews

- Game center and family sharing support

- Privacy information and;

- Compatibility with other Apple devices

Buy and download an app

1. To buy an app, tap the price. If the app is free, tap Get.

   If you there is a download icon ⬇ instead of a price, it means you have already purchased the app and you can always download the app again without necessarily paying a price.

2. You can be asked to further complete your purchase on the App store by authenticating your Apple ID with Face ID.

While your app is downloading, the app icon will show on your Home Screen together with a progress indicator. You can as well use the App library to navigate the app; the app will usually be available under the **Recently Added** Category.

Share or give an app

1. Click on the app to view its details.

2. Select ⬆, then pick a sharing option or click on Gift App (the Gift App option is not available for all of the apps you see on the App Store).

Redeem or send an App Store & iTunes gift card

1. At the top right of the app screen, simply click on 👤 or tap on your profile picture.

2. Choose from one of the listed options below;

- Redeem Gift Card or Code

- Send Gift Card by Email

Note: Using the Apple Store requires a secured internet connection and an Apple ID. The contents on the app stores are available based on geographical location; this means that some app store contents available in country A might not be available in country B. In fact, it is not all the Apple Arcade you see that will be available with service availability.

The Camera APP

Take photos with your iPhone camera

The iPhone Camera app can be used to capture a number of amazing photos and videos. There are many camera modes to select from, like the Pano mode, Video, Photo, Time-lapse, Slow-mo and portrait.

You can also use Siri to launch the Camera app.

Take a photo

The standard mode that you see first on opening your device's camera is the Photo. The Photo mode can be used to capture still photos. You will be able to select a different camera mode (like Time-lapse, Slo-mo, Pano, video and Portrait).

1. Tap on the Camera app on your Home screen or simply swipe to the left from your Lock screen to launch the Camera in Photo mode.

2. Take your photo by tapping on the shutter button. Alternatively, you can press either the volume up or volume down button to capture the shot.

Turn the flash ON or OFF

- On your iPhone camera, tap on ⚡ to turn on/off the flash. Or, click on ⌃, and then tap on ⚡ below the camera frame to select On, Auto or Off.

- On iPhone X and earlier, tap ⚡, then choose Auto, On, or Off.

Set a timer

- On your device camera, click on ⌃, and then tap ⏱.

- On iPhone X and earlier, tap ⏱.

Zoom in or out

- Launch your camera app and pinch your screen to Zoom in or Zoom out.

Take a selfie

Use your front camera to capture a selfie in Portrait or Photo mode.

1. Tap on ⟳ or ⟳ as the case may be to switch to your device's front camera.

2. Hold the iPhone 12 in front of you.

   Tip: You can tap on the arrows inside the camera's frame if you want to increase your field of view.

3. Tap on the Shutter button, or press any of the volume down or volume up button to take your shot.

To shoot a mirrored selfie, one that can captures your shot just as you see the shot in your camera frame, simply navigate to Settings ⚙, tap on "Camera," and then activate the "Mirror Front Camera."

Adjust the camera's focus and exposure

Before you capture a photo, your device camera will automatically set the exposure and focus, and the face detection will normally balance the exposure across many faces. Carry out the following actions to adjust the exposure and focus manually;

1. Tap on the camera screen to bring exposure settings and the automatic focus area.

2. Tap on anywhere on the screen where you wish to move the focus area to.

3. Next to the camera's focus area, simply adjust the exposure by dragging ☀ up or down.

   To lock the manual focus and the exposure settings for your upcoming shots, simply touch and then hold the camera's focus area until you can access the AE/AF Lock and then tap on the screen to unlock the settings.

The exposure for the upcoming shot can be set and locked. On your device, click on ⌃ , tap ⊕, and then increase or decrease the slider to adjust exposure. The exposure will remain locked till anytime you launch your Camera app. To prevent the exposure control from resetting anytime you launch your Camera app, simply navigate to the Settings app ⚙, tap on Camera, click on "Preserve Settings," and then activate the "Exposure Adjustment" by turning it on.

Take low-light photos with Night mode

On later iPhone versions, the Night mode will help capture more information and then brighten the shots when you are in a low light condition. Although the length of the exposure in **Night mode** is determined automatically, there is a chance to adjust the length manually.

On your iPhone 12, the Night mode is located on your phone's front camera for selfies, on your Ultra Wide (0.5x) camera, and also on your Wide (1x) camera.

1. Select Photo mode. In a low-light situation, the Night mode will turn on automatically: you will see the ⬤ button at the top of

your screen turning yellow and there will be a number next to the button to show how many seconds your phone camera will take to capture.

2. You can experiment with the Night mode by tapping on and then utilize the slider right below the frame to select between the Max and Auto timers. With Auto timers, the time will be determined automatically; Max deploys the longest time. The setting you make here will be preserved for the next Night mode shot

3. Tap on the camera's shutter button and then hold your camera still to capture your shot.

You will see crosshairs in the frame if the device notices movement during capture — you can simply align the crosshairs if you want to limit motions and improve the quality of your shot.

To cease capturing a Night mode shot mid-capture, simply click on the Stop button below the slider.

Take a Live Photo

Use the Live Photo to capture what is really going on just before you capture your shots and after you must have captured your shot, including the sound.

1. Select the "Photo mode."

2. Click on ⊚ to turn on/off Live Photos.

3. Click on the camera's shutter button to capture the shot.

The loop and bounce effect can be added to Live Photos.

Take a panorama photo

The Pano mode can be used to shoot landscapes or any other shots that, ordinarily, will not fit on the camera screen.

1. Select Pano mode and then tap on the Shutter button.

2. Slowly pan slowly in the arrow's direction, while keeping the arrow on the centerline.

3. Click on the shutter button once more to finish.

You will be able to pan in the opposite direction by tapping on the arrow. Rotate your device to the landscape mode to pan vertically. The direction of a vertical pan can also be reversed.

Take a photo with a filter

1. Select portrait mode or Photo, click on, and then tap.

2. Below the viewer, swipe the filters to the right or left to preview the filter and then tap on one of them to choose.

Take Burst shots

Burst Mode is when your device camera takes a series of images in rapid succession, at about ten frames/seconds. The burst shot mode provides a good means of shooting an unexpected event or an action scene. You will be able to take Burst shots by using your device's front and rear camera.

1. On your camera app, simply swipe your Shutter button to the left side to shoot rapid-fire photos. The counter is indicative of how many shots you have actually recorded.

2. Remove your finger to cease taking pictures in burst shots.

3. To mark the pictures that you wish to keep, click on the Burst thumbnail and then choose "Select." Gray dots right below the thumbnails mark the suggested pictures to keep.

4. Click on the circle in the lower-right side of each picture that you want to save as a single photo and then click on **"Done."**

To permanently delete the entire Burst, simply tap on the thumbnail and then click on 🗑.

Tip: You can take Burst shot with the Volume up button by pressing and holding on the button. Navigate to the **Settings** app , tap on **Camer**a, and then activate the "Use Volume Up for Burst."

Take videos with your iPhone 12 camera

Use the Camera app on your device to take videos. You can also change the shooting mode to slow-motion or time-lapse videos. For privacy purpose, you will see a green dot at the top right side of the camera while using the camera.

Record a video

1. Select Video mode.

2. Click on the Record button or alternatively press either the volume up or volume down button to begin recording. The following actions can be taken even while recording a video;

- Snap a still picture by pressing the white Shutter button.

- Zoom in and out by Pinching the screen.

- To get a more definite zoom, touch and then hold the 1x and then adjust by dragging the slider to the left.

2. Click on the Record button or alternatively press either the volume up or volume down button to cease recording.

The video on your iPhone 12, by default, records at 30 frames per seconds (fps). You can select other frame rates by navigating to **settings,** tap on **Camera** and select **"Record Video."** You should know that the faster your fps and the higher the resolution, the larger your video file.

Stereo sound is achieved by the iPhone 12 by using multiple microphones. You can disable the Stereo recording by navigating to "Settings ⚙," tap on "Camera," and then disable Record Stereo Sound.

The iPhone 12 will usually record videos in HDR and can also share the video with users using devices running on iOS 13.4, macOS 10.15.4 and iPadOS 13.4; while other devices will receive the same video as an SDR version. You can disable the HDR recording by navigating to **Settings** ⚙, tap on "**Camera**," select "**Record Video**," and then disable the HDR Video.

Use quick toggles to change video resolution and frame rate

In Video mode, utilize the quick toggles located at the top of your screen to change your video resolution and frame rates on your device.

On your device, tap on the quick toggles located in the top-right side to change between the 4K or HD recording and 24, 30, or 60 frames per second (fps) in the Video mode.

Record a QuickTake video

A Quicktake video is the one taken in Photo mode. You can still adjust the record button into lock position to keep capturing still pictures even while you are recording a QuickTake video.

1. In Photo mode, simply touch and then hold the Camera's Shutter button to initiate taking a QuickTake video.

2. Do a hands-free recording by sliding your shutter button to the right side and leave the lock. You will see the Shutter button and the Record button right below the frame. You can take a still picture while recording b tapping on the Shutter button.

3. To cease recording, click on the Record button.

Tip: Press and hold either your volume up or volume down button to initiate taking a QuickTake video in Photo mode.

Click on the thumbnail to see the QuickTake video in your Photos app.

Record a slow-motion video

When you take a video in Slo-mo mode, the video will record as a normal video and the slow-motion will only be seen when you play the video back. The video can as well be edited to let the slo-motion begin and ends at a time you indicate.

1. Select Slo-mo mode. Record in slow-motion on iPhone 12 with your front camera by tapping on .

2. Click on the Record button or alternatively press either the volume up or volume down button to begin recording. You will still be able to shoot a still photo while recording by tapping on the shutter button.

3. Click on the Record button or alternatively press either the volume up or volume down button to cease recording.

To set a portion of your video to play in a slow motion and the rest of the video to play at regular speed, click on the video thumbnail, and then select "Edit." You can then slide the vertical bars just below the frame viewer to choose the section you wish to play back in a slow motion.

If you want to change the slow-mo settings, navigate to **Settings**, tap on **Camera** and select "**Record Slo-mo**."

Capture a time-lapse video

Record footages at chosen intervals to make a time-lapse video. To create a time-lapse video;

1. Select Time-lapse mode.

2. Set up the iPhone 12 exactly where you plan to record a scene in motion.

3. Start recording by tapping on the Record button; tap on the Record button again to cease recording.

Tip: For enhanced results in iPhone 12, you can deploy a tripod to make your time-lapse videos. This helps to capture more brightness and details in low-light conditions.

Adjust Auto FPS settings

iPhone 12 can reduce the frame rate to 24 frame rate per seconds (fps) to improve the quality of the video taken in low light situations. To do this,

Navigate to **Settings**, tap on **Camera**, select "**Record Video**," click on **"Auto FPS,"** and then apply Auto FPS to only 30 frame rate per seconds (fps) video or to both the 30 fps and 60 fps.

Take Portrait mode photos with your iPhone 12 cameras

Use a depth-of-field effect that can keep objects or people you are shooting very sharp while at the same time creating a blurred background.

Take a photo in Portrait mode

The studio-quality lightning effect can be applied to photos in portrait mode on your iPhone 12.

1. Select Portrait mode.

2. Adjust the object properly inside the yellow portrait box by following the tips that will be provided on the screen.

3. Drag the box to select a lighting effect:

   - *Natural Light:* The face will be in sharp focus against a background that is blurred.

- *Studio Light:* The face will be brightly lit, and your photo will have a clean look.

- *Contour Light:* The face will have a dramatic shadow having highlights and lowlights.

- *Stage Light:* The face will be spotlit against a background that is deep.

- *Stage Light Mono:* The photo will be in classic black and white. The effect is just similar to the one in Stage light.

- *High-Key Light Mono:* Makes a grayscale subject on a background that is white.

2. Take the picture by tapping on the shutter button.

The Phone App

Make a call on iPhone 12

The Phone App on the iPhone is used to send a call. To send a call, simply enter the phone number on the keypad. You can also select a number to call from the contact list.

How to dial a number

Use Siri. Just say "dial" or "call" and then follow it by the number you want to call. For quick response, try and say each number in the phone digit separately.

If you don't want to use Sir, you can dial a number with the following methods;

1. Click on Keypad.

2. Carry out any of the actions below;

- *Use a different line:* On your device, click on the line at the top to choose a line if you have more than one SIM (iPhone 12 is Dual SIM).

- *Enter the number using the keypad:* Just click on the ✕ if there is any mistake while entering the number you want to call on the keypad.

- *Redial the last number:* Tap on the 📞 to view the last phone number you called or dialled, and then click on the 📞 to send a call to the number.

- *Paste a number you have copied:* Tap on the phone number field above your keypad for about three seconds (without lifting your finger), and then select **Paste.**

- *Enter a soft (2-second) pause:* Touch and then hold the star (*) key until you see a comma.

- *Enter a hard pause (if you want to pause dialling until you click on the Dial button):* Touch and then hold the (#) key until you see a semicolon.

- *Enter a "+" for international calls:* Place your finger on the "0" key until you see the "+" sign.

2. Tap on the ☎ to send the call. You can also terminate the call by tapping on the ☎.

Call your favorites

1. Click on Favorites, and then select one favorite to send a call to. Since iPhone 12 is a dual SIM device, it is capable of selecting lines for your call in the following ways;

   - If you have set a preferred line for calls, iPhone will use that line to send the call.

   - iPhone can also use the line you used to receive the last call from the contact or the line you used to send the last call to the contact.

   - The default line for voice.

2. You can manage your Favorites list by doing any of these;

- *Add a favorite:* Click on the ✛, and then select a contact.

- *Rearrange or delete favorites:* Select **"Edit."**

Redial or return a recent call

Use Siri. You can voice out something like: "Redial the number I called last" or "Return the last call."

You can as well carry out any of the action below;

1. Tap on **Recents**, and then select one number to send a call.

2. To access more details about a particular call and the call sender, simply click on ⓘ.

    A red badge will usually indicate the actual number of calls missed.

  Send a call to someone on your Contacts list

Use Siri. You can say something like: "Call David's mobile." Or do any of the following:

1. In your Phone app, select "Contacts."

2. Tap on the contact, and then click on the phone number that you want to send a call to. The call will usually be sent with the default voice line if you have not set any preferred line for that particular contact you want to call.

Change your outgoing call settings

1. Navigate to the Settings app on your device and tap on Phone.

2. Perform any of these actions;

   - *Turn on Show My Caller ID:* (GSM) Find your phone number in My Number. For your FaceTime calls, the phone number is shown automatically even if you have disabled the Caller ID.

   - *Turn on Dial Assist for international calls:* (GSM) when you turn on the "Dial Assist", your iPhone will automatically add the correct local prefix or international prefix anytime you send calls across to your favorites and contacts.

How to send emergency calls on iPhone 12

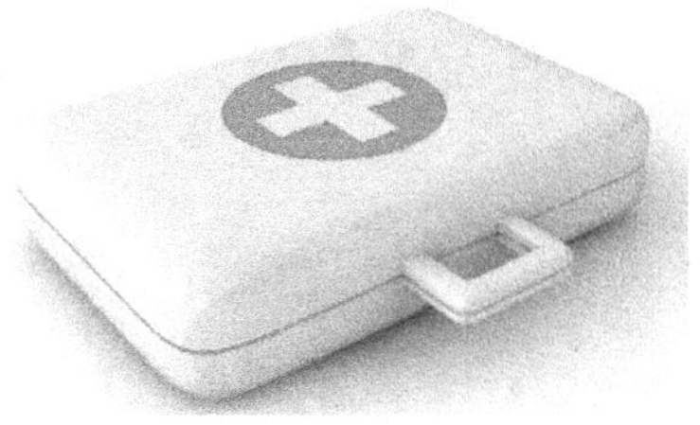

In case of any emergency, you can use your device to call for help. With the Emergency SOS, you will be able to easily call for help and notify contacts on your emergency list.

If you are sharing your Medical ID, your medical information can be sent by your device to the emergency services when you text or call 911 (the United States only).

Dial the emergency number when iPhone is locked

1. On the Lock screen, click on Emergency.

2. Enter the emergency number (for instance, 911 in the United States.), and then click 📞.

Use Emergency SOS (all countries or regions except India)

On your iPhone 12, press and then hold your side button together with the volume up or volume down button. When you see the Emergency SOS, do not lift your hands from the buttons until your device brings a warning sound and begins a countdown (you can skip the countdown by dragging the Emergency SOS slider). As soon as the countdown stops, your device will send a call across to the emergency services.

Or, you can even allow your device to initiate Emergency SOS anytime you tap on your side button five times. Navigate to **Settings**, tap on "Emergency SOS," and then activate the "**Call with Side Button.**"

Once the emergency call has been sent to the right contact, your device will notify the emergency contacts about your current location, if it is available.

How to answer or decline incoming calls on iPhone

Any incoming call can be, if you want, silenced, answered or declined. Once you decline an incoming call, the call will enter voicemail.

Answer a call

You can answer your call by doing any of these;

- Tap on .

- If your iPhone is locked, you can drag the slider.

Silence a call

Use the side button to silence a call. Once you silence a call, the call will no longer ring out. You can also use either the volume up button or the volume up button to silence a call. You will still be able to respond to a call you have silenced before it finally enters voicemail.

Decline a call and send it directly to voicemail

Carry out any of these actions;

- Press your phone two times, albeit quickly.

- Click on .

- You can also decline a call by swiping up on the call banner.

To access more options, swipe down on the call banner.

CHAPTER FIVE

CarPlay

Introduction to CarPlay

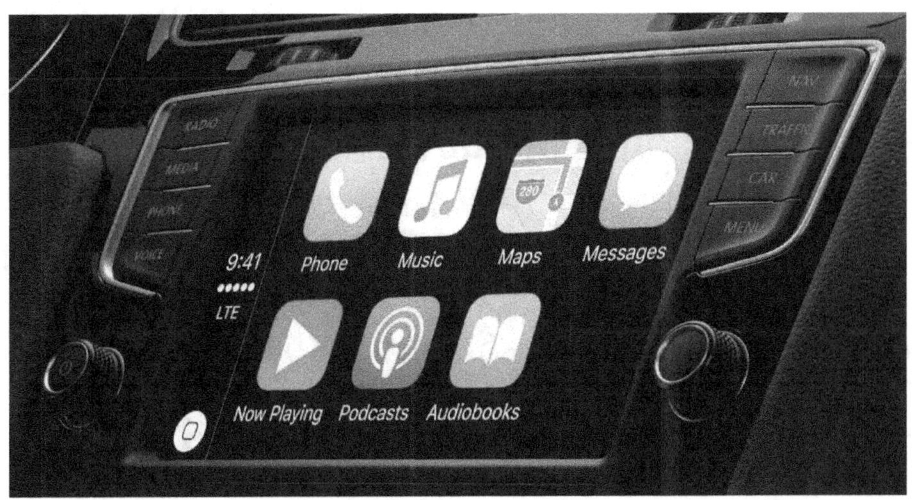

Get directions, send calls, check your calendar, listen to music and do more with CarPlay by connecting your iPhone to CarPlay. CarPlay is now present as a feature in some cars as car manufacturers have started adopting Apple innovation in their car models.

Connecting iPhone to CarPlay

Use your car's USB port or the car wireless capability to connect your vehicle and your iPhone. When you have successfully connected your car and iPhone, you can then proceed to set up your iPhone.

Before you proceed to connect the iPhone to your vehicle, you need to make sure that Siri is activated on your device. If you have not yet activated Siri on iPhone, simply navigate to "Settings , tap on "Siri & Search," and then turn on the "Press Side button for Siri."

How to connect using USB

Use a lightning to USB cable approved by Apple to connect your iPhone to the car's USB port.

A car play logo or an iPhone image can be used to label the car's USB port.

Connect CarPlay wirelessly

1. If you have a car or vehicle that support the wireless CarPlay, simply do any of these;

   - On your car's steering wheel, press and then hold the car's voice command button.

   - Ensure that your car is in Bluetooth pairing mode or wireless.

2. On your device (iPhone 12), navigate to **Settings**, tap on **"General,"** select **CarPlay** and click on **"Available Cars."**

3. Select your own car or vehicle.

Note: Some vehicles or cars that allow wireless CarPlay will let you pair by simply plugging your iPhone into the USB port of your vehicle, with a Lightning to USB cable. Once you have successfully connected an iPhone with a vehicle, you will be asked if you wish to pair wireless CarPlay for future purpose. If you choose yes, your iPhone will automatically connect wirelessly to CarPlay anytime you are driving. On some car models, the CarPlay Home will automatically appear anytime you connect your iPhone. However, in case you don't see the CarPlay Home, choose the logo of the CarPlay on the display of your car or vehicle.

Using Siri to control CarPlay (use Siri to ask CarPlay to do things)

Ask Siri on CarPlay

1. Carry out any of the following activities until you see Siri bleeps;

   - On your steering wheel, press and hold your voice command button.

   - Touch and then hold your CarPlay's Dashboard ▓≡ or the CarPlay Home ▦ button on the touchscreen showing CarPlay.

2. Ask a question or prompt Siri to do things for you.

Ask Siri. You can say;

- "Get me directions to the nearest post office"

- "Call James"

- "Play songs from Selena Gomez"

- "What is today's weather?"

- "Remind me to call David when I get home."

Use CarPlay to get turn-by-turn directions

Get turn-by-turn direction, total time of travel and travelling conditions by using Siri. Your iPhone 12 must be connected to the internet and its location service turned on before you will be able to get directions.

Find a route

CarPlay gives you possible destinations by deploying addresses from your text messages, email, calendars and contacts—as well as locations you frequent. You will also be able to search for any location, find nearby places and also use places you saved as favorites.

Ask Siri. You can say something like:

- "Take me home"

- "Give me directions to the nearest gas station"

Follow turn-by-turn directions

The CarPlay is able to voice out turn-by-turn direction to wherever you are going. You can carry out the any of the actions below while travelling;

- *End directions at any time:* You can say things like Siri "Stop navigating," or choose the estimated time of arrival (ETA) display located at the lower left, and then select End Route.

- *Mute turn-by-turn directions:* Choose the ETA display, and then select Mute.

About the Author

Derrick Richard tech expert with incredible expertise in computer diagnosis. He loves reviewing latest gadgets and has written several user-friendly manuals for Samsung and Apple devices. He passionately follows latest tech trends and his passion is in figuring out the solution to complex problems.

Derrick holds a Bachelor and a Master's Degree in ICT respectively from Georgetown University, Washington DC. He lives in Sarasota, Florida.